Book Cover by Jeremy Marriott with autumn tree background by *Ammann*/Adobe Stock Photos. Illustrations by Jeremy Marriott with autumn tree background by *Ammann*/Adobe Stock Photos.

ISBN: 978-1-7644800-9-3

1st edition 2026

Contact the Author at admin@insightarchitecturalconsulting.com.au

Visit the Author's website at https://www.insightarchitecturalconsulting.com.au/

Reference list:

American Psychiatric Association. (2022). Diagnostic and statistical manual of mental disorders (5th ed., text rev.). https://doi.org/10.1176/appi.books.9780890425787

CONTENTS

Dedication..........

Preface..........

Arch. Plan Land Lot Geometries: AUTISM (Defined)&ADHD (Undefined)..........1

ADHD: Land Lot Definitions x 18..........2

ADHD: Lots #1, #2, #7 & #8 - Geometries..........3

Lots #1, #2, #7 & #8..........4

- Misses out on details resulting in errors.
- Misplaces important things.
- Finds it very hard to concentrate.
- Is easily distracted by people & noises.

DESIGN GUIDE 1 - CENTRALLY LOCATED STORAGE
DESIGN GUIDE 2 - SOCIAL ACOUSTICS

ADHD: Lots #3, #4, #5, #6 & #9 - Geometries..........5

Lots #3, #4, #5, #6 & #9..........6

- Misses important instructions.
- Finds it very hard to organise things.
- Shy's away from starting tasks that need consistent focus.
- Finds it very hard to remember important things.
- Finds it very hard to finish tasks once begun.

DESIGN GUIDE 1 - REMINDER DISPLAYS
DESIGN GUIDE 2 - SCROLLING LED VIS.DISPLAY – AUDIOREMINDER

ADHD: Lots #10, #12, #14 & #17 - Geometries..........7

Lots #10, #12, #14 & #17..........8

- Seems to be always fidgeting, tapping, and not sitting still.
- Need to physically move causes them and others irritation.
- Physically moving as if driven by a fast internal motor.
- Need to move prevents waiting in lines or taking-turns.

DESIGN GUIDE 1 - FIDGET ENABLED SEATING
DESIGN GUIDE 2 - CREATIVE STUDIO

CONTENTS

ADHD: Lot #11 - Geometry........9

Lot #11: Unwillingly switches focus from one task to another........10

DESIGN GUIDE 1 - SPACES FOR FOCUS
DESIGN GUIDE 2 - ROOM OF SINGULAR FOCUS

ADHD: Lots #13, #15, #16 & #18 - Geometries........11

Lots #13, #15, #16 & #18........12

- Social interactions can be overly loud.
- Always seems to rush when engaged in social conversation.
- Speaks a lot and often.
- Interrupts when others are speaking socially.

DESIGN GUIDE 1 - ACOUSTICALLY MODIFIABLE SOCIAL SPACES
DESIGN GUIDE 2 - SOCIAL FLOOR INDICATORS

Arch. Plan Land Lot Geometries: AUTISM (Def.) & INTELLECTUAL DISABILITY (Undef.)..........13

ID: Land Lot Definitions x 3........14

ID: Lot #1 - Geometry & A differently functioning brain........15&16

DESIGN GUIDE 1 - VISUAL WAYFINDING
DESIGN GUIDE 2 - SENSORY MAXIMUS

ID: Lot #2 - Geometry & Difficulties adapting to change........17&18

DESIGN GUIDE 1 - HOME BASE 4 RELATED OBJECTS
DESIGN GUIDE 2 - THERMAL AUTOMATION

ID: Lot #3 - Geometry & Lots #1 and #2 are persistent from early in life....19&20

DESIGN GUIDE 1 - C-DECK: MEDICAL POD
DESIGN GUIDE 2 - HISTORICALLY REPEATING

Architectural Plan Land Lot Geometries: AUTISM, ADHD, & ID (Defined)........21

SUMMARY........22

THE AUTHOR........23

PHOTOGRAPH, AND THE PUBLISHER........24-25

Book Series Site Plan Geometries: *Complete*........26

DEDICATION

I am inspired by the inspirational: People with Autism living their best lives in a largely non-autistic world. If only the built environment showed more consideration for their needs. I deem this highly needful. As such, by way of contribution, I offer this architectural design guide book. As you read this book and subsequent series, please allow the content to spark a perpetual interest for designing built environments that show an acceptance of neurodiversity, a movement I hereby define as: *Acceptance Architecture*. From this point forward, I extend my absolute best wishes to you for better architectural outcomes for people with Autism and those who share my aforementioned inspiration.

PREFACE

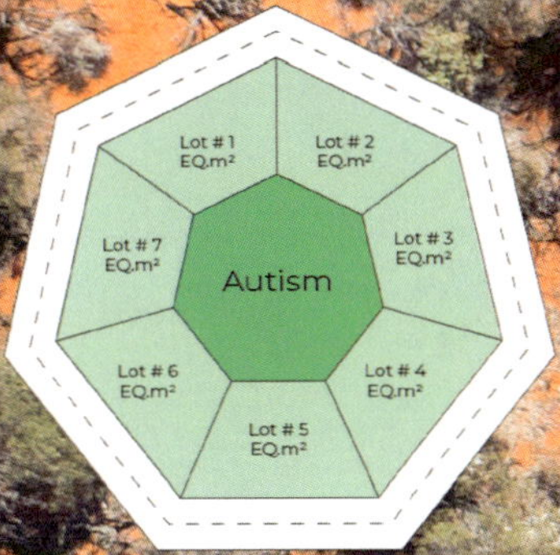

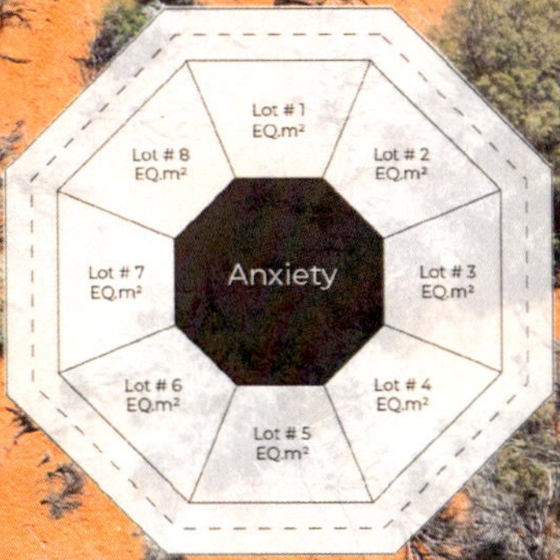

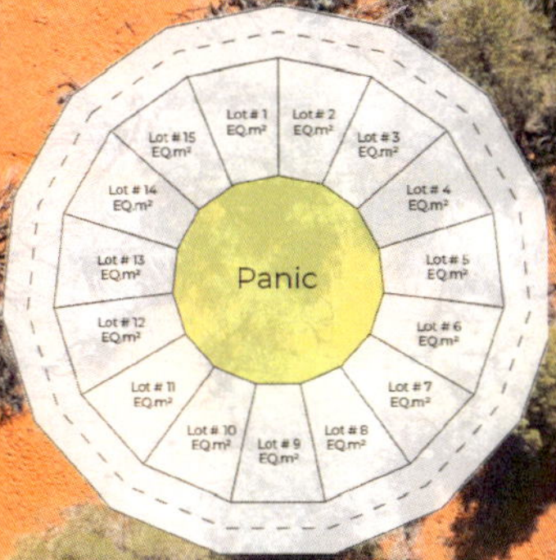

The numbered land lots within these architectural plan geometries symbolically represent and define the attributes of their central diagnoses (representative of their origin within the Diagnostic and Statistical Manual of Mental Disorders, Fifth Edition, Text Revision [DSM-5 TR, 2022]). They also form the rationale for two (2) design guidances per lot which can be used for the design of future/current built environments for people with Autism and ADHD/Intellectual Disability (ID) who identify with those particular land lot attributes. The remaining books within this series apply the other geometries (Anxiety, Depression, and Panic) to Autism as they can co-occur as is indicative of the complexity and spectrum of Autism.

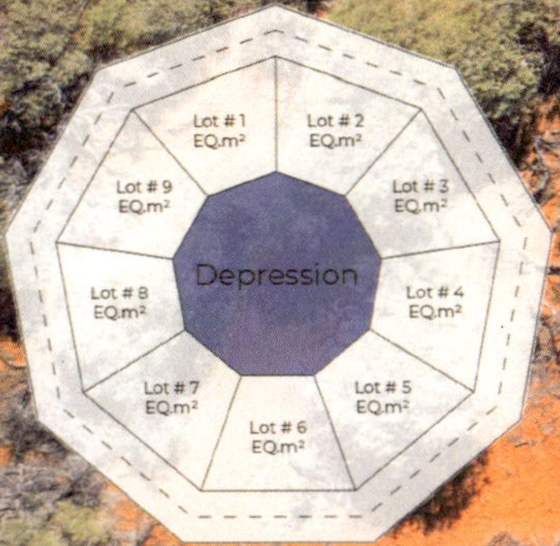

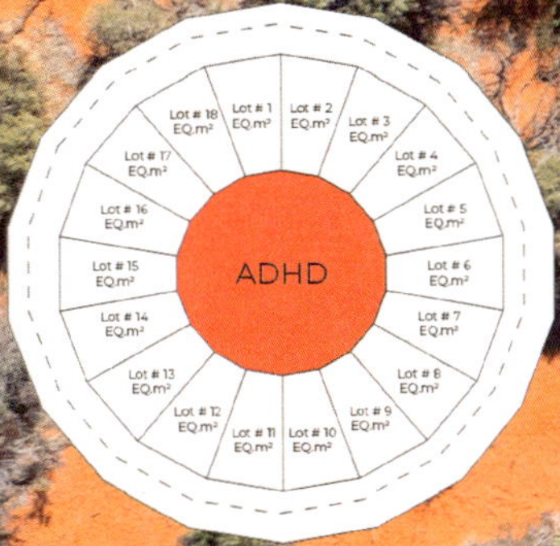

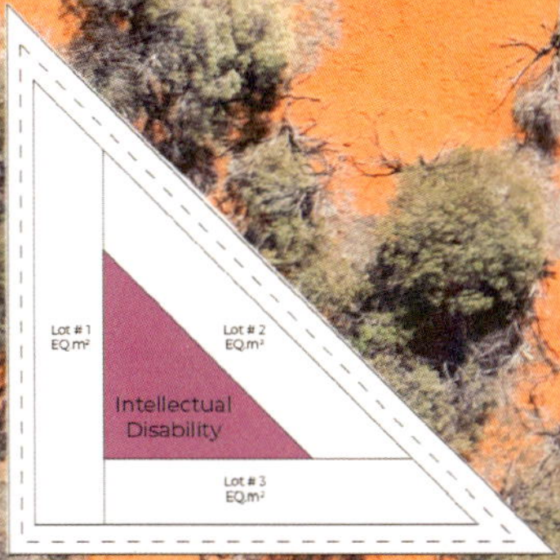

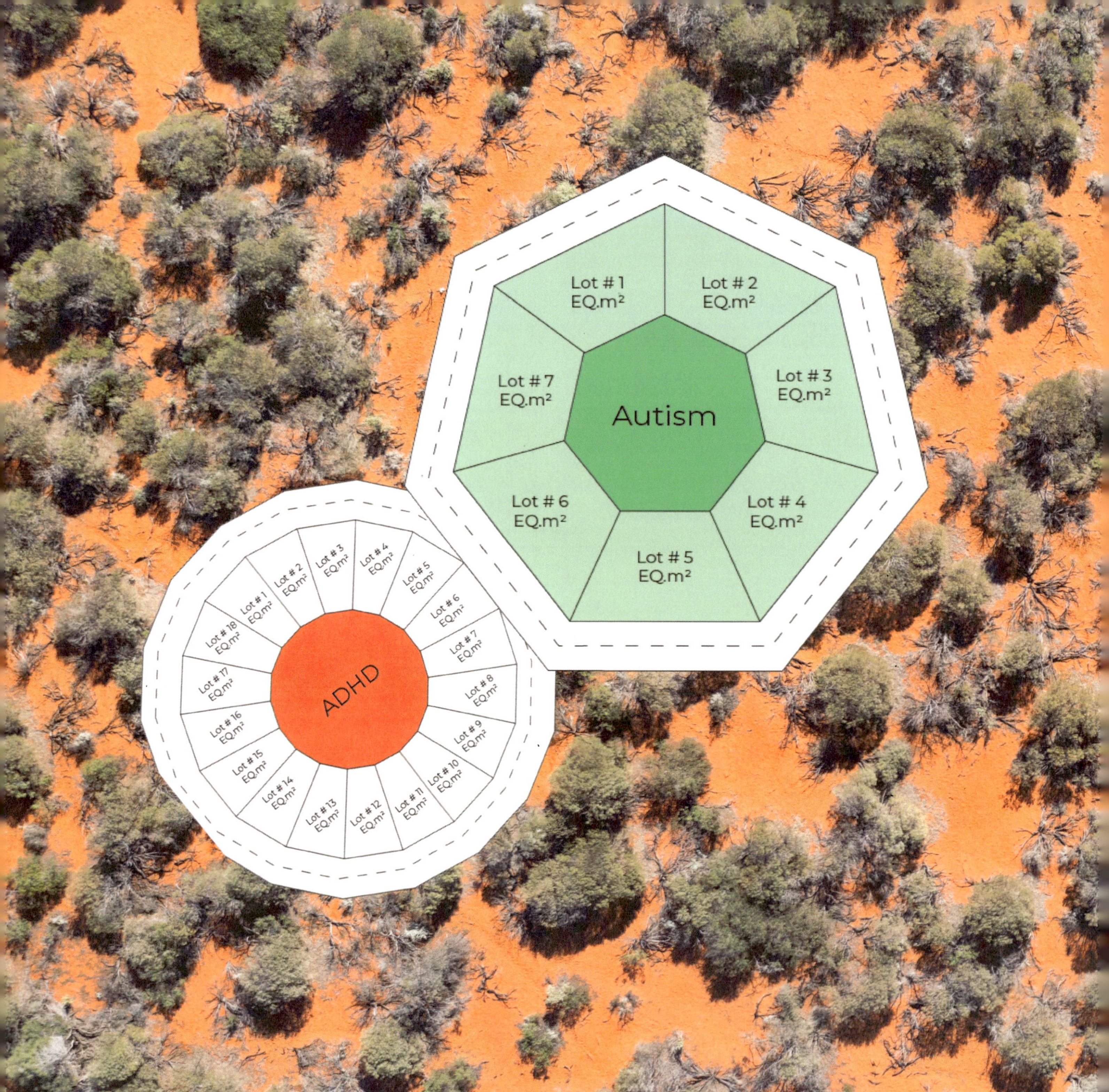
Autism
Lot # 1 EQ.m²
Lot # 2 EQ.m²
Lot # 3 EQ.m²
Lot # 4 EQ.m²
Lot # 5 EQ.m²
Lot # 6 EQ.m²
Lot # 7 EQ.m²
ADHD
Lot # 1 EQ.m²
Lot # 2 EQ.m²
Lot # 3 EQ.m²
Lot # 4 EQ.m²
Lot # 5 EQ.m²
Lot # 6 EQ.m²
Lot # 7 EQ.m²
Lot # 8 EQ.m²
Lot # 9 EQ.m²
Lot # 10 EQ.m²
Lot # 11 EQ.m²
Lot # 12 EQ.m²
Lot # 13 EQ.m²
Lot # 14 EQ.m²
Lot # 15 EQ.m²
Lot # 16 EQ.m²
Lot # 17 EQ.m²
Lot # 18 EQ.m²

AUTISM & ADHD: LAND LOT DEFINITIONS

Land Lot #1. Misses out on details resulting in errors.

Land Lot #2. Finds it very hard to concentrate.

Land Lot #3. Misses important instructions.

Land Lot #4. Finds it very hard to finish tasks once begun.

Land Lot #5. Finds it very hard to organise things.

Land Lot #6. Shy's away from starting tasks that need consistent focus.

Land Lot #7. Misplaces important things.

Land Lot #8. Is easily distracted by people and noises.

Land Lot #9. Finds it very hard to remember important things.

Land Lot #10. Seems to always be fidgeting, tapping, and not sitting still.

Land Lot #11. Unwillingly switching focus from one task to another.

Land Lot #12. Need to physically move causes them and others irritation.

Land Lot #13. Social interactions can be overly loud.

Land Lot #14. Physically moving as if driven by a fast internal motor.

Land Lot #15. Speaks a lot and often.

Land Lot #16. Always seems to rush when engaged in social conversation.

Land Lot #17. Needing to move prevents waiting in lines or taking-turns.

Land Lot #18. Interrupts when others are speaking socially.

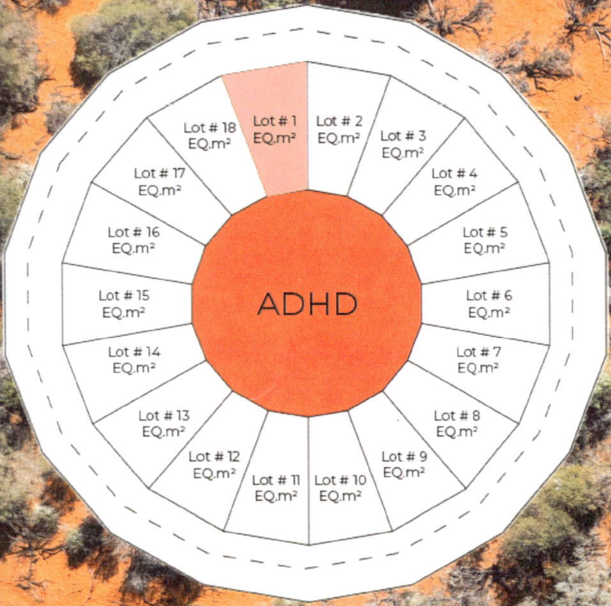

ADHD
Lot # 1 EQ.m^2
Lot # 2 EQ.m^2
Lot # 3 EQ.m^2
Lot # 4 EQ.m^2
Lot # 5 EQ.m^2
Lot # 6 EQ.m^2
Lot # 7 EQ.m^2
Lot # 8 EQ.m^2
Lot # 9 EQ.m^2
Lot # 10 EQ.m^2
Lot # 11 EQ.m^2
Lot # 12 EQ.m^2
Lot # 13 EQ.m^2
Lot # 14 EQ.m^2
Lot # 15 EQ.m^2
Lot # 16 EQ.m^2
Lot # 17 EQ.m^2
Lot # 18 EQ.m^2

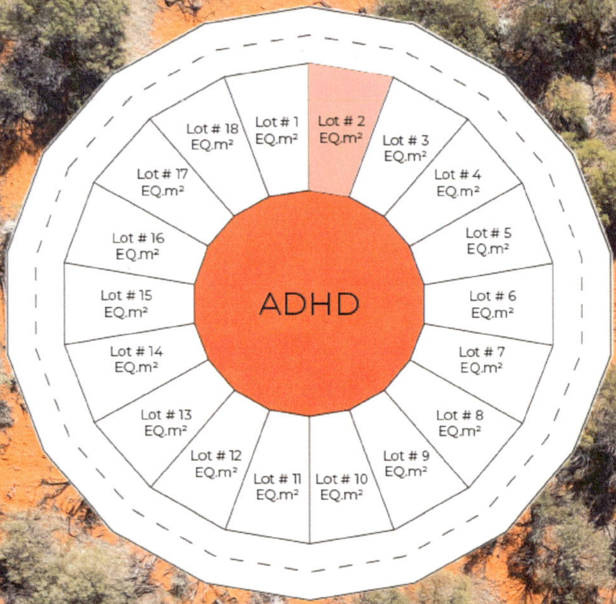

ADHD
Lot # 1 EQ.m^2
Lot # 2 EQ.m^2
Lot # 3 EQ.m^2
Lot # 4 EQ.m^2
Lot # 5 EQ.m^2
Lot # 6 EQ.m^2
Lot # 7 EQ.m^2
Lot # 8 EQ.m^2
Lot # 9 EQ.m^2
Lot # 10 EQ.m^2
Lot # 11 EQ.m^2
Lot # 12 EQ.m^2
Lot # 13 EQ.m^2
Lot # 14 EQ.m^2
Lot # 15 EQ.m^2
Lot # 16 EQ.m^2
Lot # 17 EQ.m^2
Lot # 18 EQ.m^2

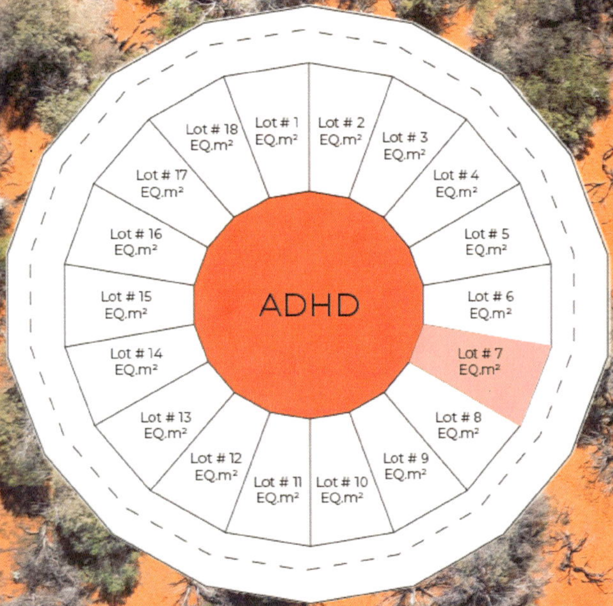

ADHD
Lot # 1 EQ.m^2
Lot # 2 EQ.m^2
Lot # 3 EQ.m^2
Lot # 4 EQ.m^2
Lot # 5 EQ.m^2
Lot # 6 EQ.m^2
Lot # 7 EQ.m^2
Lot # 8 EQ.m^2
Lot # 9 EQ.m^2
Lot # 10 EQ.m^2
Lot # 11 EQ.m^2
Lot # 12 EQ.m^2
Lot # 13 EQ.m^2
Lot # 14 EQ.m^2
Lot # 15 EQ.m^2
Lot # 16 EQ.m^2
Lot # 17 EQ.m^2
Lot # 18 EQ.m^2

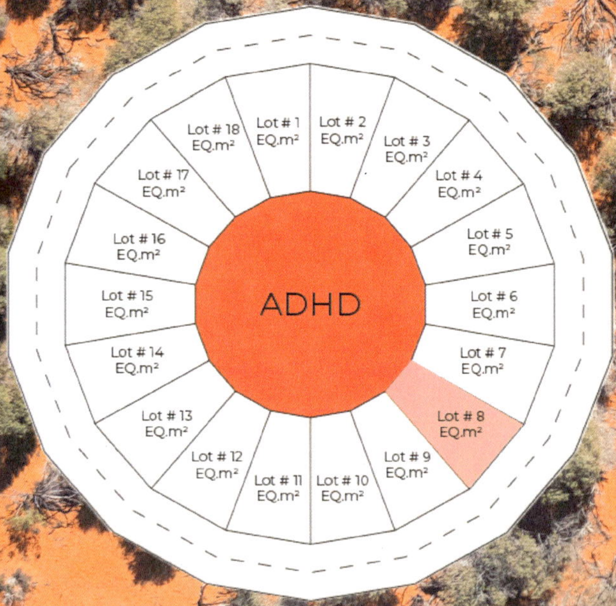

ADHD
Lot # 1 EQ.m^2
Lot # 2 EQ.m^2
Lot # 3 EQ.m^2
Lot # 4 EQ.m^2
Lot # 5 EQ.m^2
Lot # 6 EQ.m^2
Lot # 7 EQ.m^2
Lot # 8 EQ.m^2
Lot # 9 EQ.m^2
Lot # 10 EQ.m^2
Lot # 11 EQ.m^2
Lot # 12 EQ.m^2
Lot # 13 EQ.m^2
Lot # 14 EQ.m^2
Lot # 15 EQ.m^2
Lot # 16 EQ.m^2
Lot # 17 EQ.m^2
Lot # 18 EQ.m^2

LOTS #1, #2, #7, & #8

Misses out on details resulting in errors.
Finds it very hard to concentrate.
Misplaces important things.
Is easily distracted by people and noises.

DESIGN GUIDE 1 - CENTRALLY LOCATED STORAGE

CENTRALLY LOCATED STORAGE for important items (keys, TV/garage remote controls, mobile phones etc.) may be functionally beneficial for people with these attributes. Tailor-made cabinetry with see-through lockable doors fitted with key or fob unlocking mechanisms may assist in an item check-out and check-in cognitive process with the status of item(s) being linked to a digital/portable inventory display.

DESIGN GUIDE 2 - SOCIAL ACOUSTICS

Managing the volume of noise experienced in spaces used for social interactions may improve social function. As such, interior/exterior SOCIAL ACOUSTIC architectural treatments applied to places of social interactions (lounge/dining rooms, outdoor entertaining spaces etc.) are an important consideration. Examples may include surface-mounted acoustic panels, soft furnishings, outdoor hedges etc.

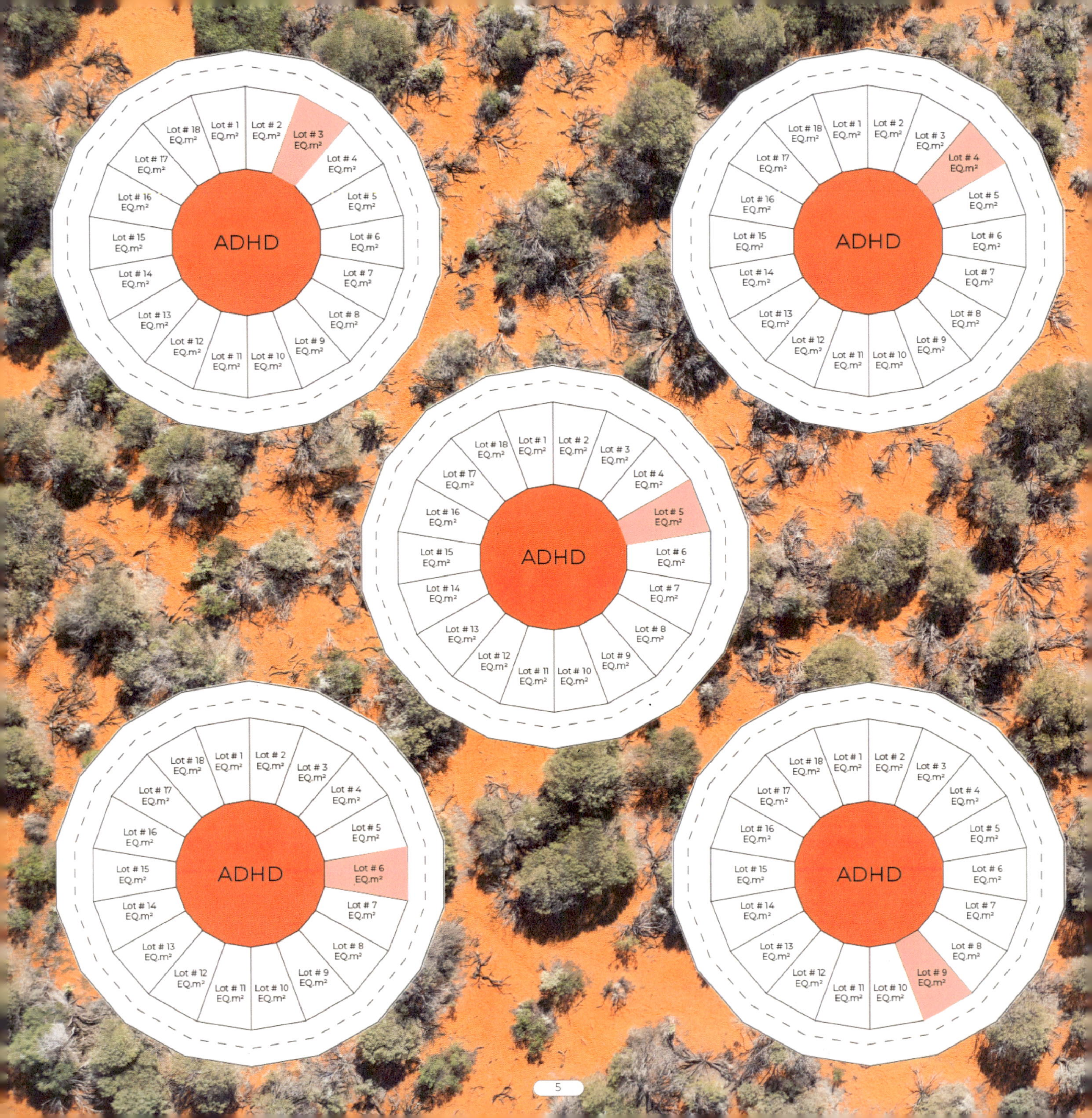
ADHD
Lot # 1 EQ.m^2
Lot # 2 EQ.m^2
Lot # 3 EQ.m^2
Lot # 4 EQ.m^2
Lot # 5 EQ.m^2
Lot # 6 EQ.m^2
Lot # 7 EQ.m^2
Lot # 8 EQ.m^2
Lot # 9 EQ.m^2
Lot # 10 EQ.m^2
Lot # 11 EQ.m^2
Lot # 12 EQ.m^2
Lot # 13 EQ.m^2
Lot # 14 EQ.m^2
Lot # 15 EQ.m^2
Lot # 16 EQ.m^2
Lot # 17 EQ.m^2
Lot # 18 EQ.m^2
ADHD
Lot # 1 EQ.m^2
Lot # 2 EQ.m^2
Lot # 3 EQ.m^2
Lot # 4 EQ.m^2
Lot # 5 EQ.m^2
Lot # 6 EQ.m^2
Lot # 7 EQ.m^2
Lot # 8 EQ.m^2
Lot # 9 EQ.m^2
Lot # 10 EQ.m^2
Lot # 11 EQ.m^2
Lot # 12 EQ.m^2
Lot # 13 EQ.m^2
Lot # 14 EQ.m^2
Lot # 15 EQ.m^2
Lot # 16 EQ.m^2
Lot # 17 EQ.m^2
Lot # 18 EQ.m^2
ADHD
Lot # 1 EQ.m^2
Lot # 2 EQ.m^2
Lot # 3 EQ.m^2
Lot # 4 EQ.m^2
Lot # 5 EQ.m^2
Lot # 6 EQ.m^2
Lot # 7 EQ.m^2
Lot # 8 EQ.m^2
Lot # 9 EQ.m^2
Lot # 10 EQ.m^2
Lot # 11 EQ.m^2
Lot # 12 EQ.m^2
Lot # 13 EQ.m^2
Lot # 14 EQ.m^2
Lot # 15 EQ.m^2
Lot # 16 EQ.m^2
Lot # 17 EQ.m^2
Lot # 18 EQ.m^2
ADHD
Lot # 1 EQ.m^2
Lot # 2 EQ.m^2
Lot # 3 EQ.m^2
Lot # 4 EQ.m^2
Lot # 5 EQ.m^2
Lot # 6 EQ.m^2
Lot # 7 EQ.m^2
Lot # 8 EQ.m^2
Lot # 9 EQ.m^2
Lot # 10 EQ.m^2
Lot # 11 EQ.m^2
Lot # 12 EQ.m^2
Lot # 13 EQ.m^2
Lot # 14 EQ.m^2
Lot # 15 EQ.m^2
Lot # 16 EQ.m^2
Lot # 17 EQ.m^2
Lot # 18 EQ.m^2
ADHD
Lot # 1 EQ.m^2
Lot # 2 EQ.m^2
Lot # 3 EQ.m^2
Lot # 4 EQ.m^2
Lot # 5 EQ.m^2
Lot # 6 EQ.m^2
Lot # 7 EQ.m^2
Lot # 8 EQ.m^2
Lot # 9 EQ.m^2
Lot # 10 EQ.m^2
Lot # 11 EQ.m^2
Lot # 12 EQ.m^2
Lot # 13 EQ.m^2
Lot # 14 EQ.m^2
Lot # 15 EQ.m^2
Lot # 16 EQ.m^2
Lot # 17 EQ.m^2
Lot # 18 EQ.m^2

LOTS #3, #4, #5, #6, & #9

Misses important instructions.
Finds it very hard to finish tasks once begun.
Finds it very hard to organise things.
Shy's away from starting tasks that need consistent focus.
Finds it very hard to remember important things.

DESIGN GUIDE 1 - REMINDER DISPLAYS

REMINDER DISPLAYS populated with written/digital/symbolic directions/ task prompts located along a wayfinding path may assist by prompting recall, reducing attentional drift, and/or promoting task completion. These (multiple) displays may be located (possibly wall-mounted) at main wayfinding junctions/nodes for ease of reference with content being self-populated and/or supported if needed as per LAND LOT #5 of the Autism symbolic geometry (BOOK 1).

DESIGN GUIDE 2 - SCROLLING LED VISUAL DISPLAY – AUDIOREMINDER FUNCTION

As a possible addition to LOT #5: Needing routines, schedules, & familiarities - DESIGN GUIDE 2 - SCROLLING LED VISUAL DISPLAY from BOOK 1, an AUDIOREMINDER FUNCTION may be included to alert people with these attributes to reference the content of the REMINDER DISPLAYS above and/ or prompt a memory recall attempt.

ADHD
Lot # 1 EQ.m^2
Lot # 2 EQ.m^2
Lot # 3 EQ.m^2
Lot # 4 EQ.m^2
Lot # 5 EQ.m^2
Lot # 6 EQ.m^2
Lot # 7 EQ.m^2
Lot # 8 EQ.m^2
Lot # 9 EQ.m^2
Lot # 10 EQ.m^2
Lot # 11 EQ.m^2
Lot # 12 EQ.m^2
Lot # 13 EQ.m^2
Lot # 14 EQ.m^2
Lot # 15 EQ.m^2
Lot # 16 EQ.m^2
Lot # 17 EQ.m^2
Lot # 18 EQ.m^2
ADHD
Lot # 1 EQ.m^2
Lot # 2 EQ.m^2
Lot # 3 EQ.m^2
Lot # 4 EQ.m^2
Lot # 5 EQ.m^2
Lot # 6 EQ.m^2
Lot # 7 EQ.m^2
Lot # 8 EQ.m^2
Lot # 9 EQ.m^2
Lot # 10 EQ.m^2
Lot # 11 EQ.m^2
Lot # 12 EQ.m^2
Lot # 13 EQ.m^2
Lot # 14 EQ.m^2
Lot # 15 EQ.m^2
Lot # 16 EQ.m^2
Lot # 17 EQ.m^2
Lot # 18 EQ.m^2
ADHD
Lot # 1 EQ.m^2
Lot # 2 EQ.m^2
Lot # 3 EQ.m^2
Lot # 4 EQ.m^2
Lot # 5 EQ.m^2
Lot # 6 EQ.m^2
Lot # 7 EQ.m^2
Lot # 8 EQ.m^2
Lot # 9 EQ.m^2
Lot # 10 EQ.m^2
Lot # 11 EQ.m^2
Lot # 12 EQ.m^2
Lot # 13 EQ.m^2
Lot # 14 EQ.m^2
Lot # 15 EQ.m^2
Lot # 16 EQ.m^2
Lot # 17 EQ.m^2
Lot # 18 EQ.m^2
ADHD
Lot # 1 EQ.m^2
Lot # 2 EQ.m^2
Lot # 3 EQ.m^2
Lot # 4 EQ.m^2
Lot # 5 EQ.m^2
Lot # 6 EQ.m^2
Lot # 7 EQ.m^2
Lot # 8 EQ.m^2
Lot # 9 EQ.m^2
Lot # 10 EQ.m^2
Lot # 11 EQ.m^2
Lot # 12 EQ.m^2
Lot # 13 EQ.m^2
Lot # 14 EQ.m^2
Lot # 15 EQ.m^2
Lot # 16 EQ.m^2
Lot # 17 EQ.m^2
Lot # 18 EQ.m^2

LOTS #10, #12, #14, & #17

Seems to always be fidgeting, tapping, and not sitting still.
Need to physically move causes them and others irritation.
Physically moving as if driven by a fast internal motor.
Needing to move prevents waiting in lines or taking-turns.

DESIGN GUIDE 1 - FIDGET-ENABLED SEATING

Indicating a designers acceptance of human attributes, people with these features may be availed opportunities to access FIDGET-ENABLED SEATING. Specifically, silent-of-use fidgeting devices may be embedded within the designs of interior/exterior seating of sufficient robustness to handle high repetition of use. Venues for such seating may include places where sitting for lengthy periods is expected (cinemas, dentists, dining rooms, classrooms etc.). Permanently affixed or clip-on fidget devices may be appropriate.

DESIGN GUIDE 2 - CREATIVE STUDIO

To meet and satisfy high-levels of physical/mental hyperactivity (especially on occasions when pharmacological treatments have washed-out), a 24/7 accessible CREATIVE STUDIO may be useful for channeling these attributes into creative pursuits and possibly SIAs as per LOT #1 DESIGN GUIDE 1: SIA IMMERSION from BOOK 3.

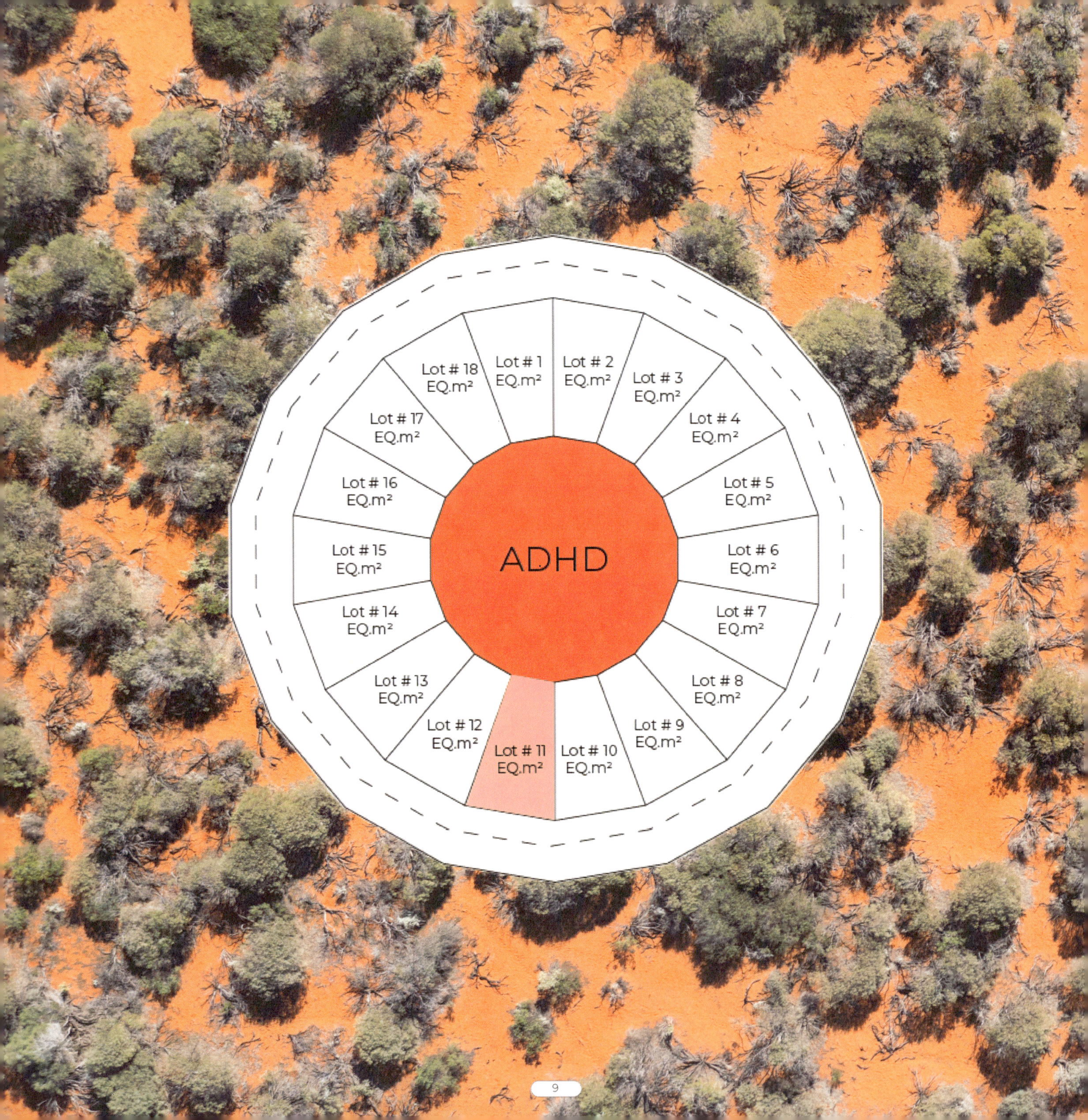
ADHD
Lot # 1 EQ.m²
Lot # 2 EQ.m²
Lot # 3 EQ.m²
Lot # 4 EQ.m²
Lot # 5 EQ.m²
Lot # 6 EQ.m²
Lot # 7 EQ.m²
Lot # 8 EQ.m²
Lot # 9 EQ.m²
Lot # 10 EQ.m²
Lot # 11 EQ.m²
Lot # 12 EQ.m²
Lot # 13 EQ.m²
Lot # 14 EQ.m²
Lot # 15 EQ.m²
Lot # 16 EQ.m²
Lot # 17 EQ.m²
Lot # 18 EQ.m²

LOT #11

Unwillingly switching focus from one task to another.

DESIGN GUIDE 1 - SPACES FOR FOCUS

The very nature of this attribute means that people may transition unwillingly from being focused to unfocused resulting in task derailment. As such, specific SPACES FOR FOCUS (partitioned, floor marked, or enclosed rooms) devoid of distracting elements may be beneficial. Including essential furniture/fixings only may be useful.

DESIGN GUIDE 2 - ROOM OF SINGULAR FOCUS

People with this attribute can experience a derailment of focus due to environmental distractions. As such, a ROOM OF SINGULAR FOCUS that includes elements only associated with a particular task may be beneficial. To enable diversity of use, if possible, this room may also be designed for adaptability to accommodate for other singular uses. If so, a lockable cabinet (located outside of the room) may be included for the categorised storage of items specific to singular focus.

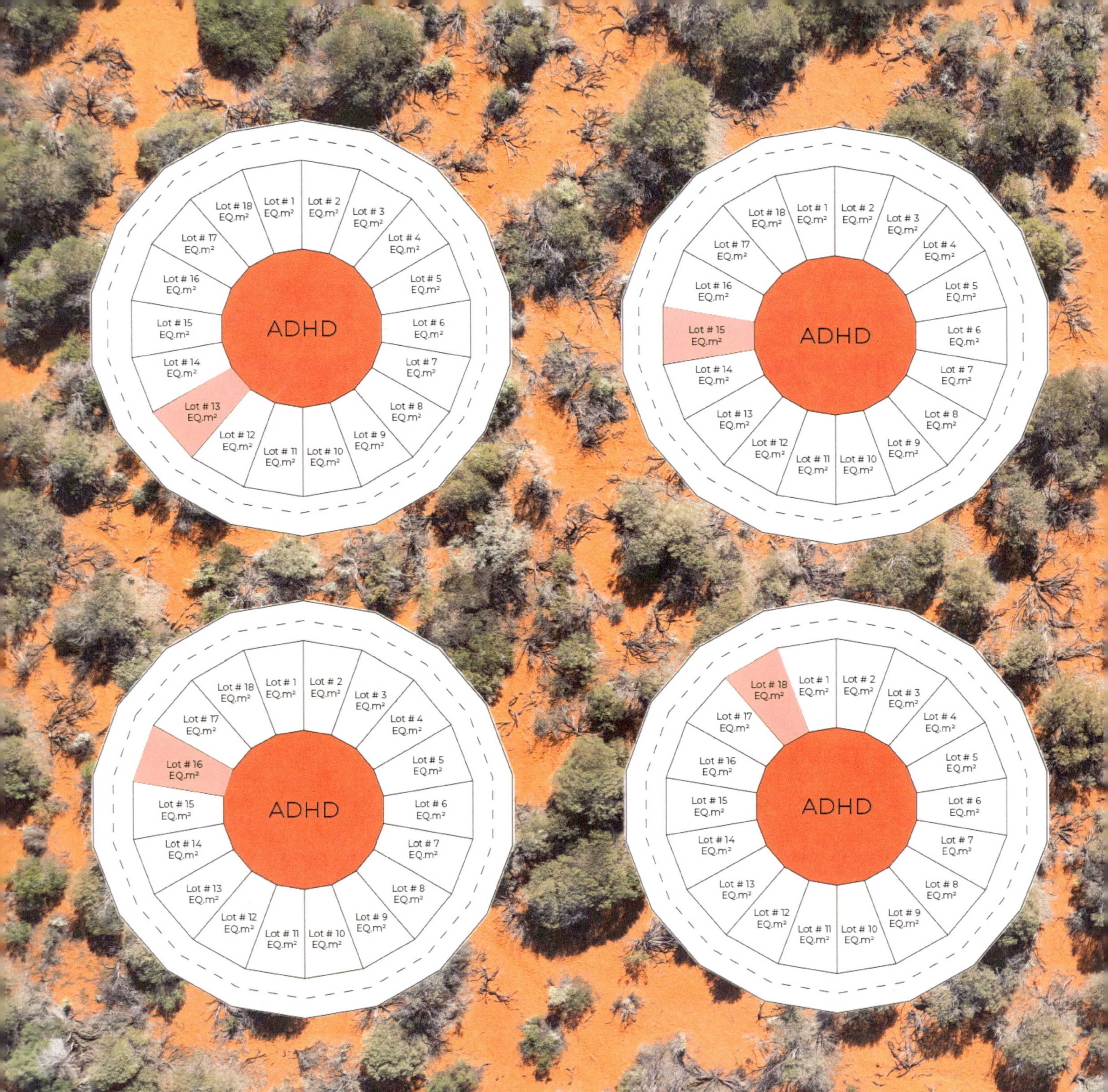

ADHD
Lot # 1 EQ.m²
Lot # 2 EQ.m²
Lot # 3 EQ.m²
Lot # 4 EQ.m²
Lot # 5 EQ.m²
Lot # 6 EQ.m²
Lot # 7 EQ.m²
Lot # 8 EQ.m²
Lot # 9 EQ.m²
Lot # 10 EQ.m²
Lot # 11 EQ.m²
Lot # 12 EQ.m²
Lot # 13 EQ.m²
Lot # 14 EQ.m²
Lot # 15 EQ.m²
Lot # 16 EQ.m²
Lot # 17 EQ.m²
Lot # 18 EQ.m²
ADHD
Lot # 1 EQ.m²
Lot # 2 EQ.m²
Lot # 3 EQ.m²
Lot # 4 EQ.m²
Lot # 5 EQ.m²
Lot # 6 EQ.m²
Lot # 7 EQ.m²
Lot # 8 EQ.m²
Lot # 9 EQ.m²
Lot # 10 EQ.m²
Lot # 11 EQ.m²
Lot # 12 EQ.m²
Lot # 13 EQ.m²
Lot # 14 EQ.m²
Lot # 15 EQ.m²
Lot # 16 EQ.m²
Lot # 17 EQ.m²
Lot # 18 EQ.m²
ADHD
Lot # 1 EQ.m²
Lot # 2 EQ.m²
Lot # 3 EQ.m²
Lot # 4 EQ.m²
Lot # 5 EQ.m²
Lot # 6 EQ.m²
Lot # 7 EQ.m²
Lot # 8 EQ.m²
Lot # 9 EQ.m²
Lot # 10 EQ.m²
Lot # 11 EQ.m²
Lot # 12 EQ.m²
Lot # 13 EQ.m²
Lot # 14 EQ.m²
Lot # 15 EQ.m²
Lot # 16 EQ.m²
Lot # 17 EQ.m²
Lot # 18 EQ.m²
ADHD
Lot # 1 EQ.m²
Lot # 2 EQ.m²
Lot # 3 EQ.m²
Lot # 4 EQ.m²
Lot # 5 EQ.m²
Lot # 6 EQ.m²
Lot # 7 EQ.m²
Lot # 8 EQ.m²
Lot # 9 EQ.m²
Lot # 10 EQ.m²
Lot # 11 EQ.m²
Lot # 12 EQ.m²
Lot # 13 EQ.m²
Lot # 14 EQ.m²
Lot # 15 EQ.m²
Lot # 16 EQ.m²
Lot # 17 EQ.m²
Lot # 18 EQ.m²

LOTS #13, #15, #16, & #18

Social interactions can be overly loud.
Speaks a lot and often.
Always seems to rush when engaged in social conversation.
Interrupts socially when others are speaking.

DESIGN GUIDE 1 - ACOUSTICALLY MODIFIABLE SOCIAL SPACES

Further to Page 4., DESIGN GUIDE 2: SOCIAL ACOUSTIC architectural treatments, people with these attributes & their social communicative partners may benefit from limiting auditory distractions by including ACOUSTICALLY MODIFIABLE SOCIAL SPACES. Removable acoustic wall-panels & retractable curtains are examples of such modifications.

DESIGN GUIDE 2 - SOCIAL FLOOR INDICATORS

People with these attributes may find social reciprocity (turn-taking) challenging due to distractibility's and non-verbal social skill difficulties (See BOOK 1, Lots #1 & #2). As such, the inclusion of SOCIAL FLOOR INDICATORS marking socially appropriate physical locations/distances for both the person with Autism/ADHD and their social communicative partner(s) may be beneficial. Examples may include footmarks for standing, location markers for chairs etc.

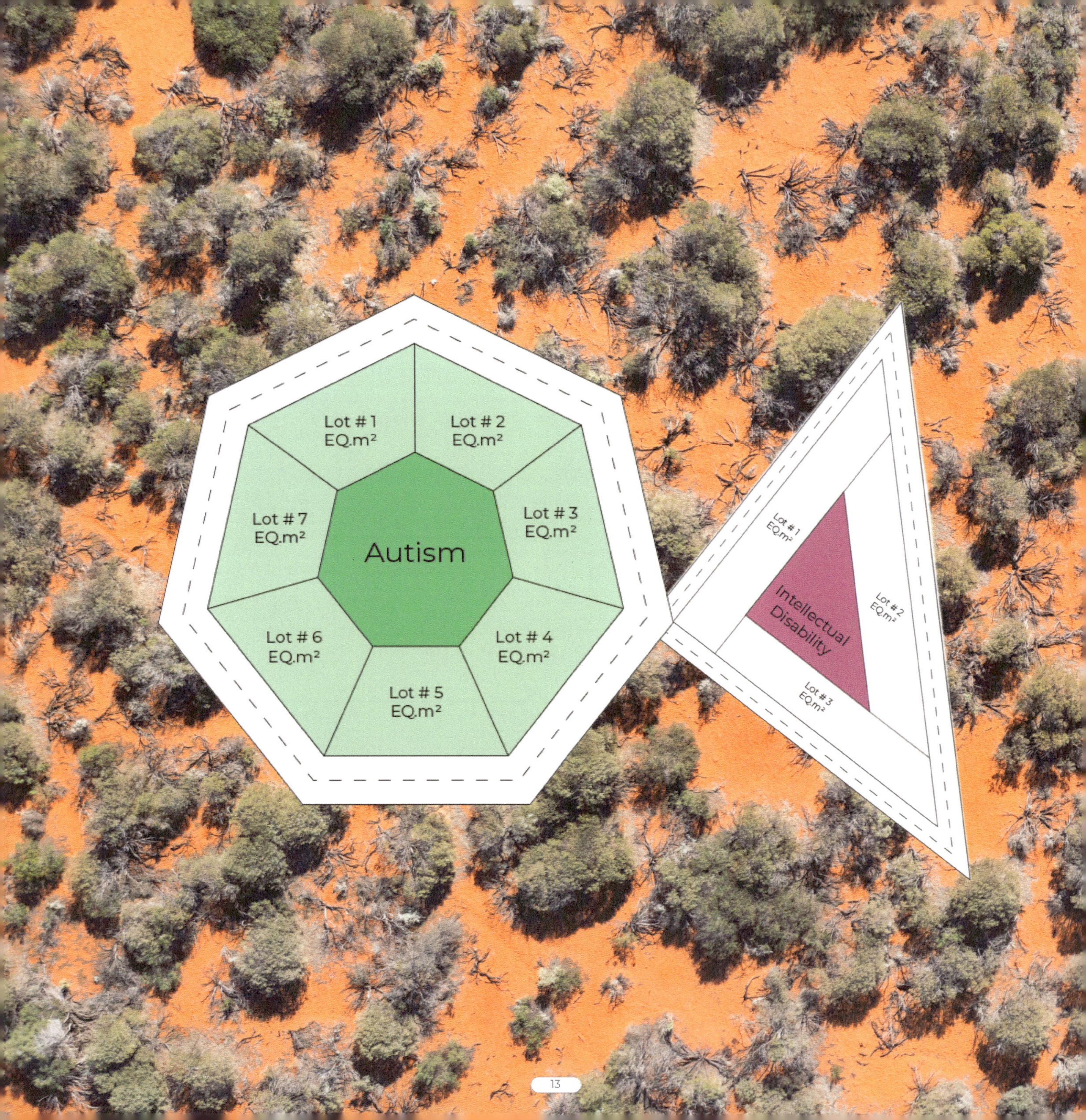
Lot # 1
EQ.m²
Lot # 2
EQ.m²
Lot # 3
EQ.m²
Lot # 4
EQ.m²
Lot # 5
EQ.m²
Lot # 6
EQ.m²
Lot # 7
EQ.m²
Autism
Lot # 1
EQ.m²
Lot # 2
EQ.m²
Lot # 3
EQ.m²
Intellectual Disability

AUTISM & ID: LAND LOT DEFINITIONS

Land Lot #1. A differently functioning brain.

Land Lot #2. Difficulties adapting to change.

Land Lot #3. Lots #1 and #2 are persistent from early in life.

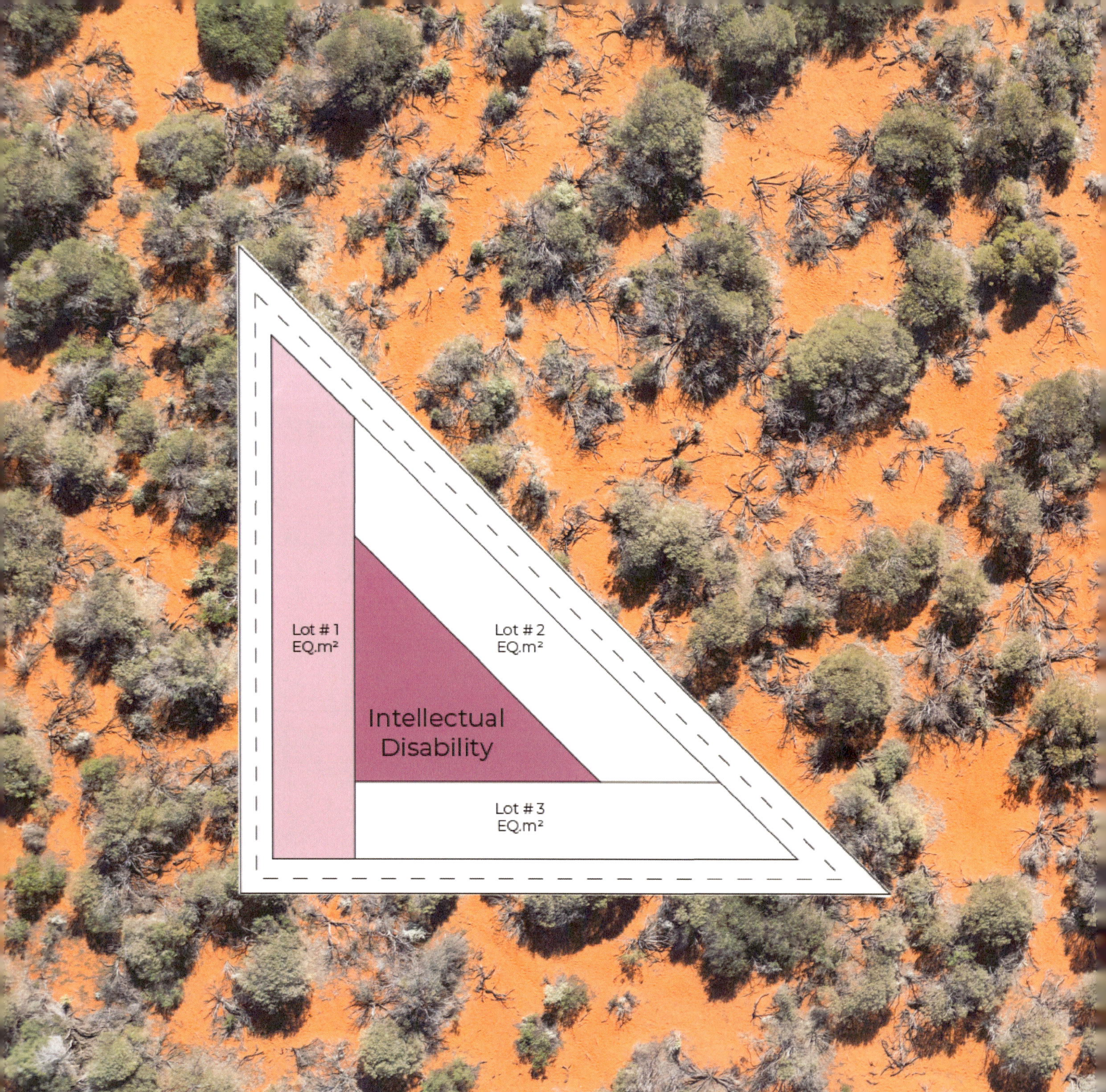
Lot # 1
EQ.m²
Lot # 2
EQ.m²
Intellectual
Disability
Lot # 3
EQ.m²

LOT #1

A differently functioning brain.

DESIGN GUIDE 1 - SIA THEMED WAYFINDING

People with this attribute may have clearly defined SIAs (as per BOOK 1, Lot #6) in addition to needing a well-thought-out wayfinding strategy. As such, SIA THEMED WAYFINDING (to attract and retain attention) may be beneficial. Wayfinding methods may include preferred objects, colours, shapes, names, sensory preferences (See BOOK 1, Lot #7), room-related items etc. This strategy may be best used in physically larger & de-cluttered pathways to limit distractions.

DESIGN GUIDE 2 - MAXIMISING SENSORY LIKES

As per BOOK 1, Lot #7, people with Autism may hold sensory dislikes & likes. These sensory attributes are especially important for people who have Autism and an ID. As such, in addition to AUTISM: Attribute - Lot#7: DESIGN GUIDE - 1 SENSORY SPACE (BOOK 1), MAXIMISING SENSORY LIKES should be a primary architectural design focus for built environments for people with Autism and an ID. Specifically, sensory spaces should be substantial/ embellished enough to enable sufficient access to sensory items/activities often.

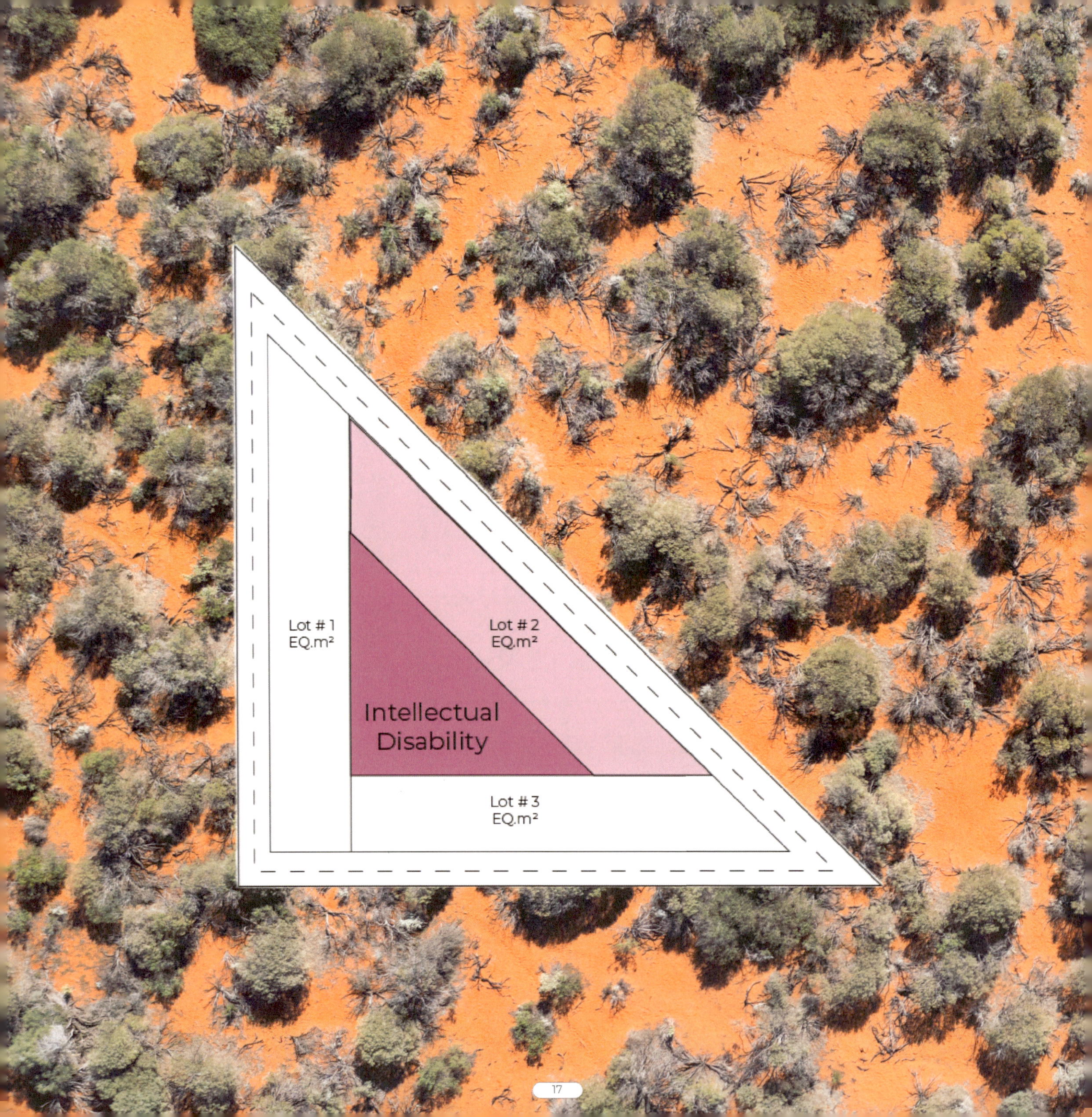
Lot # 1
EQ.m²
Lot # 2
EQ.m²
Intellectual
Disability
Lot # 3
EQ.m²

LOT #2

Difficulties adapting to change.

DESIGN GUIDE 1 - HOME BASE FOR RELATED OBJECTS

People with this attribute can experience difficulties adapting to change due to cognitive difficulties and the need for structure/routine (See BOOK 1, Lot #5). As such, a HOME BASE FOR RELATED OBJECTS (centrally located) may act as a repository for physical objects (stored in a chronological order and related to upcoming activities) that can be collected, taken to activities, then placed in a completed compartment before the collection of the next object.

DESIGN GUIDE 2 - THERMAL AUTOMATION

As published in AUTISM and ANXIETY: Lot#5: DESIGN GUIDE - 2 THERMOREGULATION (BOOK 2), thermoregulation can be vital for people with Autism. Further, people with an ID may find it difficult to self-manage their thermal comfort due to cognitive difficulties. As such, designing for automated thermal regulation may be appropriate. Mechanical and Structural methods for regulating thermal conditions may include ventilation, window shading/glazing, insulation, zoning, HVAC, thermal mass, passive cooling and heating etc.

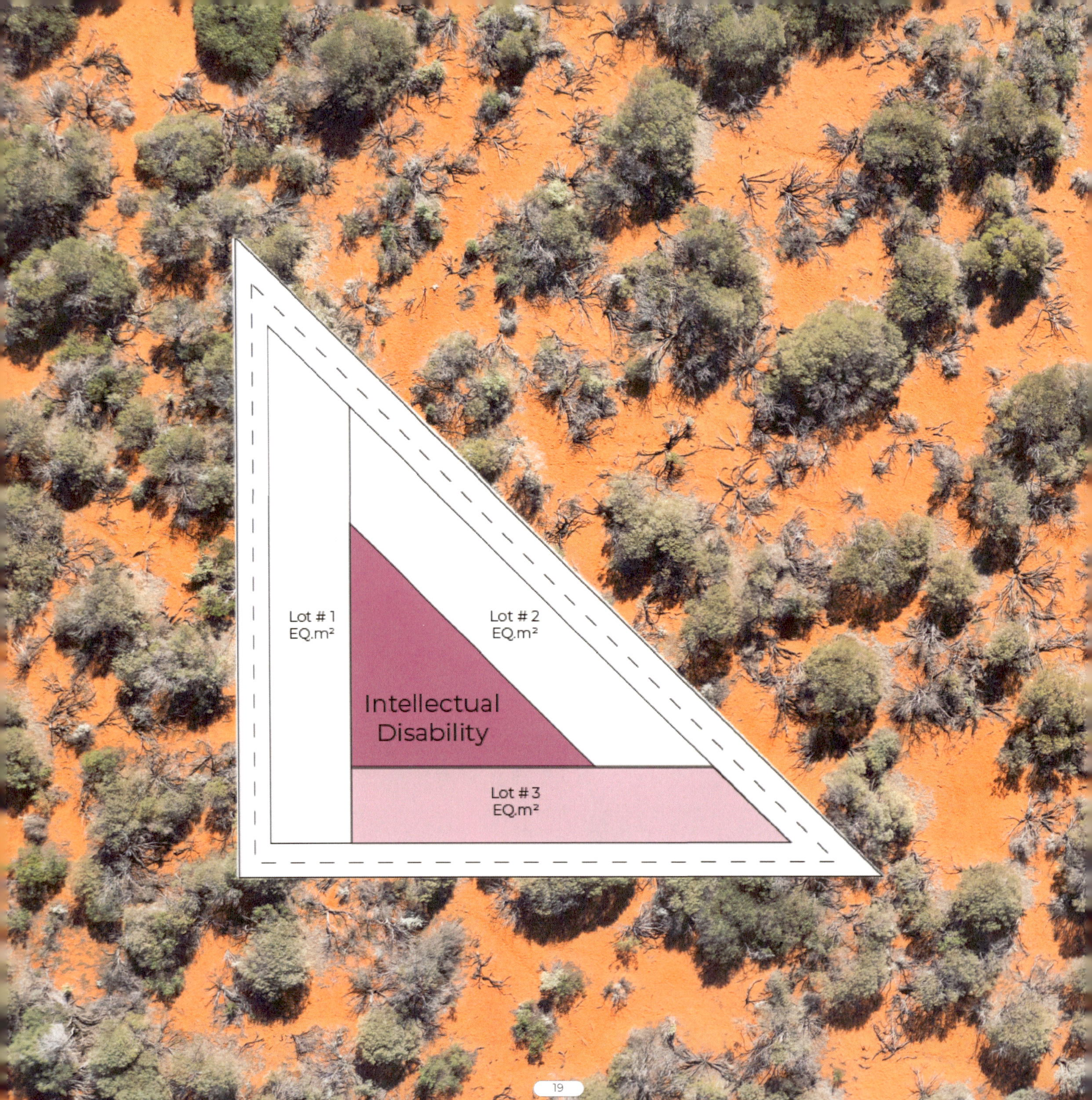
Lot # 1
EQ.m²
Lot # 2
EQ.m²
Intellectual
Disability
Lot # 3
EQ.m²

LOT #3

Lots #1 and #2 persist from early-on in life.

DESIGN GUIDE 1 - C-DECK: MEDICAL POD

As a possible addition to AUTISM: Lot#5: DESIGN GUIDE - 1 COMMAND DECK (BOOK 1), a C[ONTROL]-DECK: MEDICAL POD may provide young people with AUTISM and an ID a place for in-person or online medical and professional health service consultations in a familiar setting (aligning with BOOK 1, Lot #5). Frequent and specific medical assessment/treatment may be needed for these people due to complexity of needs. The design of the pod may accommodate for parents/primary carers and consultants possibly minimising difficulties associated with attending hospitals for medical treatment.

DESIGN GUIDE 2 - HISTORICALLY REPEATING

People with AUTISM and an ID may experience difficulties adapting to unfamiliar environments. As such, designers may consider researching past positively experienced environments, noting familiar elements, and HISTORICALLY REPEATING them in unfamiliar interior/exterior architectural settings. Examples of historical elements may include building orientation, interior furnishings, exterior elements, interior/exterior volumes etc. of past positively experiences settings.

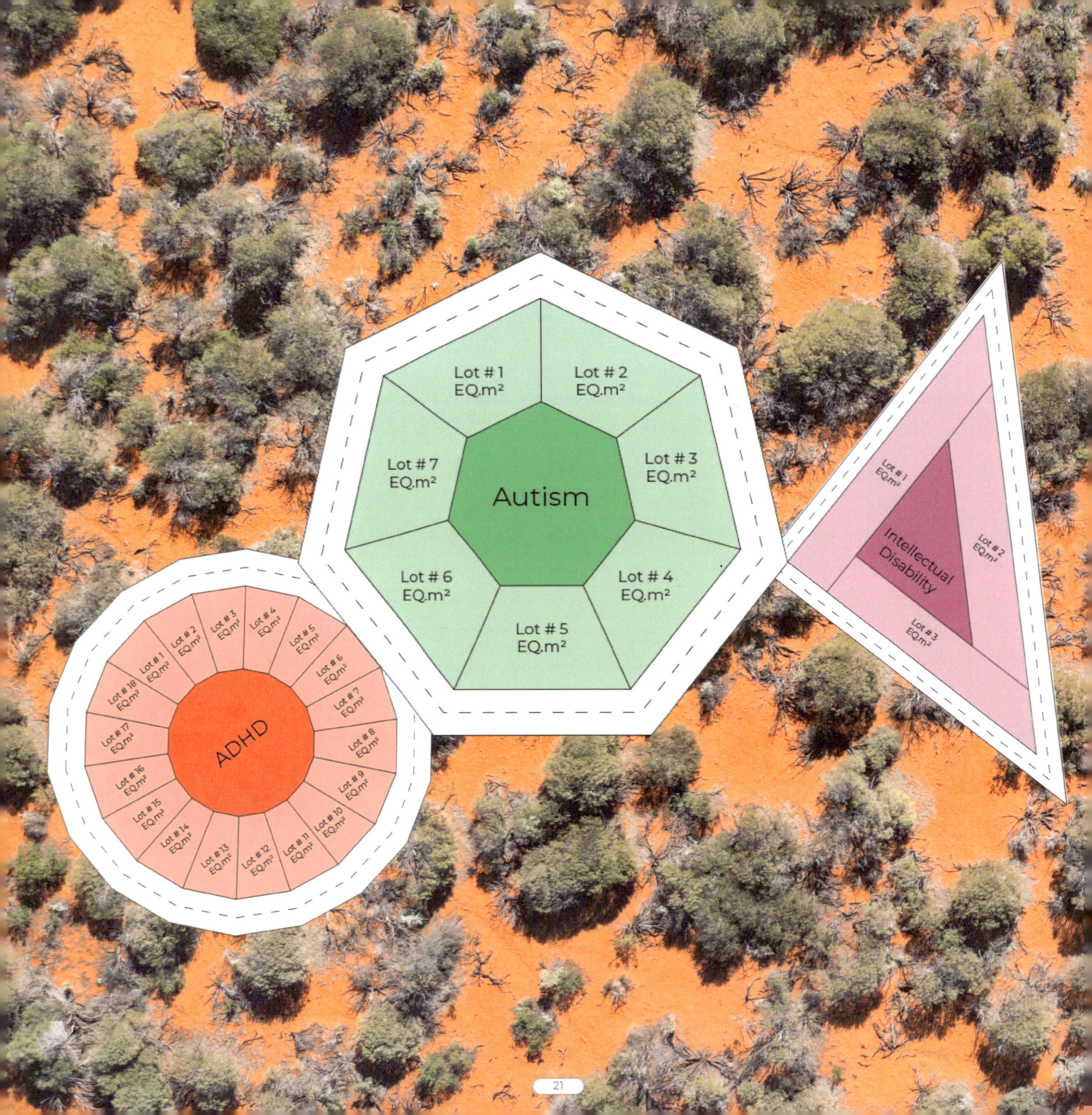
Lot # 1
EQ.m²
Lot # 2
EQ.m²
Lot # 3
EQ.m²
Lot # 4
EQ.m²
Lot # 5
EQ.m²
Lot # 6
EQ.m²
Lot # 7
EQ.m²
Autism
ADHD
Intellectual Disability
Lot # 1
EQ.m²
Lot # 2
EQ.m²
Lot # 3
EQ.m²

SUMMARY

The 4th (fourth) in a series of 4 (four) books, this architectural guide defined the attributes of ADHD and ID then provided specific design responses for the creation of highly-considered built environments as they relate to Autism. Using a novel method developed by the author, the diagnostic criteria for Autism from the DSM-5 TR (2022) were paraphrased and symbolically represented as land lots within an architectural plan geometry. Ultimately, in the '*Architecture for ...*' book series, the geometry created for Autism was considered alongside Anxiety, Depression, Panic, ADHD, and ID to provide design responses that cover the complexity and spectrum of Autism. The sum of the author's work now culminates in a unified an back-page architectural site plan geometry worth consideration now and far into the future.

THE AUTHOR

Dr Jeremy Marriott (BA Psych/Dis, BSc Psych (Hons), MPsych (Clin), PhD, BSc (Arch Sc) strives to inform and inspire humanistic architectural ventures. His writing process is an organic derivative of all he has learned and his dedication to making the world a better place for people living with neurodiversity. He hopes that his work can make positive differences in the lives of those who read, imagine, create, design, renovate, build, and live in the resultant architecture. His knack, if you will, is the ability to learn then combine disciplines for the betterment of those for whom architecture is intended: All peoples. With the utmost respect and sincerity, he offers his work as a seminal contribution to this emerging body of knowledge.

THE PUBLISHER

Founded by the author in 2025, *Insight Architectural Consulting* applies Dr Jeremy Marriott's unique and expert blend of knowledge, focus, attention-to-detail, and insight across all phases and interior/exterior/community aspects of architectural projects intent on responding to the needs of the peoples for whom they are created. Dr Marriott adds value and insights by conducting Architectural Design Reviews, producing design review reports with enlightening and functional recommendations, and presenting empirical and applied content to contribute to the architectural body of knowledge and its emerging trends. His mission is to champion and contribute to human-centered and Acceptance Architecture designs that result in the adoption of functional strategies that translate into better user experiences for all peoples; especially those with neurodiverse needs.

CONTACT:
admin@insightarchitecturalconsulting.com.au
https://www.insightarchitecturalconsulting.com.au/

Insight Architectural Consulting

www.ingramcontent.com/pod-product-compliance
Lightning Source LLC
LaVergne TN
LVRC090954210926
848482LV00009B/353

* 9 7 8 1 7 6 4 4 8 0 0 9 3 *